What's Up, Doc?
Making the Most of Your Next Doctor Visit: What Every Patient Needs to Know

What's Up, Doc?
Making the Most of Your Next Doctor Visit: What Every Patient Needs to Know

Frances Anderson-Hewitt, MD

To my mother, who is truly the wind beneath my wings. Mom, your strength and determination have made me a better person. I can only hope to be the example to my children that you have been to me. I love you.

To my father, who has truly shown vulnerability and strength through very difficult times. I thank you. I love you. I appreciate you being my first model of a man.

Contents

Introduction

Navigating today's health-care system can be amazingly complex and difficult, even if one has all the necessary tools. I can only imagine the fear and anxiety experienced when someone is ill-equipped to assist in addressing his or her health-care concerns. I want this book to serve as a personal guide, as caring and concerned advice from a good friend who also happens to be a physician.

Acknowledgements

I want to thank Raymond for being a caring, compassionate, loving and disciplined father to our children, who accepts only the best from and for them. I want to thank my siblings, Sabrina, Pamela, Breman and Afton, for their points of view and expressions of their individual personalities. Each of you is a unique spirit who adds true balance to our family.

I also want to thank my close friends— Dawne, Monica, Tamiko, LaNelle, Sandra and Meresha—for always telling me the truth. I give thanks to Natalee and Grandma Pearl for being a support to my family, being the village and helping to raise my children. I thank Dr. Susan Schooley, who has seen me as a diamond and allowed me to shine.

Part One

Before We Get Started

This is a book for patients. If you haven't already been a patient, you will be eventually. I'm a family physician. I love my job, and I love my patients. My colleagues are commendable, and my support staff is great. I'm good at my job, but I'm not perfect, and neither is your doctor. I have good days and bad days. Most of my patients are lovable, and, over the years, I have become pretty attached to them and their families. Sometimes, however, my patients are quite challenging. I have a select few patients who cause sensations of uneasiness when I see their names on my schedule. I would venture to say, Every doctor has "that patient." It may be:

- The one who always threatens legal action whenever he or she has waited too long to be seen
- The patient who is loud, obnoxious and threatening to staff.
- The child of an elderly patient with dementia who really requires assisted living care, but no one in the family is willing to acknowledge it.
- The patient who has abused his or her body with bad diet and dangerous habits (smoking, little to no exercise, risky sexual behaviors, and so forth) for twenty to thirty years, but now blames "those doctors" when health conditions develop

Or, sometimes, it's just:

- A very ill patient with fifteen or more medications and six or seven chronic medical conditions that must be managed within our twenty-minute time limit without a family member to assist

- A very angry family member who isn't willing to listen to answers for his or her questions

I have a million more scenarios, but central to them all is the patient, the friend, or the family member who generally means well, but will not allow me, the physician, to do my job.

All I really want to do is be the doctor, which, to me, means assisting patients and their families with making appropriate and necessary choices for their health care. My goal is to present the information and all options available and to help you, the patient, to make the right health-care decisions for you. My job is not to make the decisions. For the sports fans, I'm the coach; you're the owner of the team. After careful review and consideration of all options, only you know what is ultimately best for your own situation.

Being hindered in any way from doing what I'm trained to do is one of the most challenging parts of being a family physician. I enjoy the variety. No two days are the same. I treat all—rich or poor, thin or overweight, influential or not-so influential, and homeless or mansion dwellers. All patients are really no different to me on the basis of these things, and I believe that is true for most doctors. What makes patients different is the level of clinical expertise required to care for them. You want your medical condition (not personality differences between yourself and the doctor) to dictate and be the sole determinant as to whether a high-stress situation is in order for the day.

This is a book to teach you, the patient, how to make the most of your primary care visit. This visit may actually be anywhere from five minutes to twenty-five minutes of face-to-face time with your doctor. With insurance co-pays increasing and doctors being required to see more patients in less time, you want to be able to get the most from that office visit. It is important that visits to your physician are focused and efficient, yet address all of your concerns. Now, more than ever, people have neither the time nor money for multiple unnecessary office visits. I want you and your doctor to benefit from your visit, with you being a satisfied, safe, healthy patient with any chronic diseases at least under control and possibly halted from progressing. I want your doctor to have the satisfaction of knowing that he or she does make a difference and is helping to halt the effects of disease on your body.

I do not want you to become "that patient", the one described earlier who has anything less than a positive and productive interaction with his or her doctor. Doctors are regular people who have emotions. You can imagine that, when we have an extremely confrontational patient, that individual may be sabotaging his or her own health care and not even realize it. Your doctor is not going to knowingly do you any harm, but you are not going to receive the same service and care as a preferred patient if you are "that patient".

Your doctor wants the best for you, and you deserve the best possible medical care. So how do you not become that overwhelming, demanding, just-get-him-out-of-the-office ASAP patient? It's actually easy. I'm going to give you some tips on how to not become "that patient" or to get out of this position if you have noticed that you already have.

You are too important to have your health dependent upon whether you are personable. I believe that only the most emotionally mature physicians are able to treat patients in a way that is completely blind to their own apparent judgments and prejudices. We have all had experiences with maturation, the experiences of life that make us all who we are. Completely discounting these experiences would be impossible; each doctor is both a person first and then a physician second. The intrinsic humanity of your doctor should not be a hindrance to your health care.

Your relationship with your physician should remain a professional one with a level of trust and intimacy you share. You do not want a relationship that blurs professional and personal lines. This actually could be as dangerous to your care as a doctor who is not invested in your care. Your relationship with your physician should strike a good balance between the ability of the doctor to be completely objective and you being a preferred patient. I will show you how to achieve this balance. Keep reading.

I divided this book into two parts:

- The first part is a basic primer listing steps necessary in choosing the right doctor for you.

- The second part is actually a bit of a sampling of patient stories from my own practice, with each one illustrating a common ailment, its treatment and demonstrating some principles necessary for seamless communication with your physician.

I base each scenario on a true clinical situation but, of course, I have changed all identifying information to protect those involved. For some scenarios, I have combined several patients into one for the purpose of demonstration.

In this book, you will see the term "primary care physician," which refers to the doctor who cares for your routine health maintenance. See the following for examples:

- A **pediatrician** sees children from birth generally to seventeen years, although some will see older teens.

- A family physician sees patients from birth to end of life; some even see pregnant women.

- An **internal medicine** doctor treats all adults eighteen years and older, although some will see adolescents.

Taking an Office Visit from Just Okay to Great!

No, you can't be in control of someone else's day. Nor can you control the circumstances in a doctor's life surrounding your visit. Doctors experience problems, too, for example, job dissatisfaction, burnout, bankruptcy, foreclosure, divorce and marital discord, domestic violence and abuse, substance abuse, difficulties with child rearing, sick children, elder or disabled parent or sibling care, and severe personal illness. No amount of education or financial stability can shield anyone from real-life experiences. There is no way to know if your doctor may be experiencing one of these hardships. These circumstances, if present, can unfortunately impact your visit, even if only in very small and unintentional ways.

My job is to empower you with information I feel is necessary to promote a healthy relationship with your primary care physician. It takes a little preparation on your part. Remember, you are responsible for your own health and well-being. Yes, your doctor should be held to legal and ethical standards, and, yes, you do have the option of reporting your physician and even taking legal action if you feel you have a medical malpractice concern, but I can guarantee you that you want to avoid these options by not having the mishap in the first place. No amount of financial restitution or public apology will take back the error, however small and uncomfortable or large and disabling. Your best bet is to do whatever you can to avoid it. You can avoid many errors by doing your part from the beginning. Remember, the goal is a healthy you, as healthy as is medically possible.

The following are nine tips to keep office visits and your physician relationship efficient and effective:

1.   Find a qualified physician.

2.   Find a doctor you like and trust.

3.   Do your part.

4.   Respect this relationship and make it important to you.

5.   Ask for what you want, and be direct.

6.   Believe.

7.   Pursue, pursue, and pursue.

8.   Set realistic expectations.

9.   Be open-minded.

Find a Qualified Physician

When seeking a personal physician, you prefer the doctor to have two professional criteria: board certification and an active medical license.

Board certification

Take special care to find a board-certified physician, that is, someone who has passed the testing necessary to practice his or her special field of medicine. You may review a doctor's current board certification status on the American Board of Medical Specialties Web site (www.abms.org). But, a lack of board certification does not necessarily mean that he or she is not qualified to be your doctor. Check into the doctor's special circumstances. Perhaps he or she is "board eligible," Meaning the physician just completed his or her residency training (which will be discussed later), and has not yet taken the necessary test. Or perhaps he or she may have taken a long break to pursue other interests, raise children, care for ailing parents, and so forth and must take the test again to reinstate board certification.

Active medical license

Be sure the doctor has an active medical license. Each state individually licenses the physicians who apply to practice medicine in that state. Some doctors have licenses in more than one. You may look up your doctor's medical license information on your state's Web site. Every state has a state government Web site (generally "www.your state government/health license—licensing for health-care professionals") where you can check any physician, nurse, chiropractor, and other health professional to see if:

- his or her license is active
- he or she has any disciplinary action or complaints against him or her
- his or her medical license has reached its expiration date

Negative information

If you should check either your state's Web site or the board certification sites and find negative information about a doctor, I advise you to speak directly with the physician about this. Most doctors welcome the discussion to explain the circumstances surrounding any documented incident because, in general, most physicians really want to do the best they possibly can for their patients. Negative outcomes do occur and can happen with or without fault. Even when everyone involved (doctors, nurses, support staff, patients, families, and so forth) does everything right, unfortunately, bad things still happen. Do not simply dismiss a doctor because you find something you feel is damaging. Most doctors will probably be named in at least one lawsuit in their medical careers; certain specialties are More prone to legal action than others. Before you make any decisions about your medical care, take any claims or other information you have found directly to the doctor for open discussion. If he or she is unwilling to discuss this information, then it is possible that the physician is in the midst of the case and is legally bound to not discuss any aspects of it. If he or she denies that there are legal constraints to discussing involvement and still chooses not to address your concerns, then maybe this is not the physician for you.

Medical school

Do not be too focused on which medical school your doctor attended. Be very careful about attaching labels to doctors based on medical school training. Each medical school has basic training in medical sciences that all students must demonstrate mastery of on national board examinations.

You will notice the initials "MD" (allopathic physician) or "DO" (osteopathic physician) behind your doctor's name. These letters refer to the type of medical school he or she attended. Both types of doctors are well trained in traditional Western medicine, but a DO has additional training in diagnosing disease with body manipulation. Both are licensed and board certified by the states in which they practice. You may consult the American Medical Association and American Osteopathic Association Web sites for more information.

Residency

Although it is true that the basis of medical education begins in medical school, the doctor actually develops the way in which he or she practices medicine during residency training. During their residencies, doctors are able to actually engage in the practice of medicine with supervision. (Residency training will be discussed later.) Actually, you want an experience that includes exposure to multiple patients and patient care scenarios. If you live in an inner city but a doctor primarily participated in a suburban residency, he or she may or may not be acquainted with all the aspects unique to an urban practice (and vice versa for doctors trained in urban residencies).

Find a Doctor You Like and Trust

This is the first and most important tip after finding a qualified physician. After you have determined that a doctor has adequate credentials, ask yourself:

- "Do I like my doctor?"
- "Do I trust my doctor?"
- "Do I trust my doctor to make the best decisions for my health care?"
- "Do I trust my doctor to include me in making all decisions?"
- "Do I trust him or her to listen to my opinion and take my feelings and concerns into consideration?"

You will get a different feeling about each physician you meet. For example, about ten physicians are in my hospital-based practice, and we are all completely different in our styles and delivery of medical care, but our approach to disease treatment and methods for illness prevention are generally the same. I have had patients come to me because:

- They liked my style, demeanor, and sense of humor.
- They liked the way I spoke with them as if they might be a family member or a good friend.
- They could not talk to their other doctors.

I have also had patients not see me a second time because they felt my approach was too informal. Either my tone offended them, or they mistook my approach to be flippant and disrespectful. Every patient is different, and I'm not offended if someone does not want to see me any further. You must find a doctor you like and feel comfortable sharing your concerns with without reservation. Take your time; you are worth it.

I have a very straight-laced physician colleague who hardly ever smiles. He is very knowledgeable and experienced, but, some time ago, there was a decrease in his office visits and patients' general satisfaction with him. Patients were afraid to talk to him because they felt intimidated. If I were ever ill with a chronic disease, he would be the physician I would want to see, but most patients see only his personality, the outer shell.

The most knowledgeable physician in the world cannot help anyone if patients do not feel as if they can talk to him or her. As the patient, you have a right and responsibility to find a physician you can talk to. So much of your treatment depends on you and what you say to your doctor, that is, your communication. Patient history and the discussion that follows regarding any concerns probably determines 80 percent of a diagnosis. The physical exam, lab work or x-rays, and further testing only support or disprove a physician's initial theory of why you are ill. Information taken from the patient generates the initial theory. It is important that I receive the full and accurate story initially from you, the only person who can give it to me.

Be assured that your conversations with your physician are confidential. Medical records are legal documents, and, without your permission, your doctor and anyone else who comes in contact with your medical record are legally bound to not disclose any information to a third party. Only with your consent can any doctor share information about you with anyone (except law enforcement in cases of child abuse or elder abuse and, in some states, domestic violence as well). You can feel safe that your husband, wife, children, employer (even if you work in health care), or friends will not know your medical information unless you give the doctor permission to discuss it with them.

It is time to begin your research. You must have the patience and time to look until you find what you want, need, and expect. Don't let price be your sole determining factor. Yes, we all must be cognizant of how much we are spending, but, if you find what you need and want, it may be worth it to spend a little bit more, drive a little farther, and wait a little longer. Your peace of mind is worth a small sacrifice.

You may need to interview multiple physicians or go in to be seen for minor problems to gauge your interaction with a new physician. You may not get the right physician the first, second, or third time you interview one. Also, your family member's or friend's "great doctor" may or may not be the one for you. You can and should see a doctor based upon family and friends' recommendations, but keep in mind that the perfect doctor for your friend or family member may not be your ideal.

What do you want? You want:

- A doctor who listens to you and acts on your real concerns

- A doctor who will tell you the truth based upon his or her years of medical training and experience

What do you not want? You do not want:

- A doctor you can manipulate into doing anything you feel is necessary because unfounded, haphazard, and random testing can actually be dangerous to your health and cause consequences and side effects that you haven't bargained for or couldn't foresee

You want a doctor who can be objective with you and the insurance company and will offer you the most appropriate testing for your complaints. The ideal doctor does not overtest to appease you, but also does not undertest to avoid damaging the existing relationship between himself or herself and your insurance company. It may take time to find a physician who is professionally mature enough to be able to strike this balance.

Find someone you trust to be your partner and advocate for your health. Keep in mind that you must be reasonable. Only the very rich can have their own personal physicians. You are not your doctor's only patient. Although you are entitled to your doctor's full attention during your visit and follow-ups for any test results, medications, or other questions you have, you are not entitled to another fifteen- or twenty-minute telephone conversation after your visit. You may have to make another appointment to completely address further questions.

Do Your Part

It is important to write down, if you can, a list of things that are troubling you. It is also important for you to

14

understand that your doctor will probably only be able to sufficiently address the top three or four things on your list. The doctor will need to determine what those things are. You cannot. Even if you do have some medical background or are even a physician yourself, you cannot see everything your doctor sees. Your priorities may not be the same. For example, you may be concerned about your flaking scalp because of the obvious social stigma and wish treatment for this. Your doctor may be more concerned about that episode of confusion or chest pain you had last week. You may require multiple visits to address everything on your list. Please allow your doctor to determine in which order he or she will address your concerns.

If you need to return for more visits to address all of your concerns, this is not because the doctor is attempting to get more money out of you. Remember, your appointment may be as few as five minutes in length. It is very unnerving when patients pull out a list of twenty or more complaints. Your doctor cannot address all these in one visit. In my opinion, it is impossible for a doctor to evaluate more than three or four unrelated concerns fully and completely in one office visit. Sometimes, depending on the nature and severity of the problem, the threshold may be even less. For example, if you are a newly diagnosed diabetic, your office visit may be entirely devoted to this. You will need to return for your acne treatment or discussion of your constipation. It is not that your other concerns are not worthy of attention. They are. It is simply more important that you allow your doctor to determine the order in which he or she addresses your concerns. The doctor does not decide whether the concerns are ever addressed. He or she only decides the exact order of tackling each concern. No matter how many appointments it takes, you will need to be consistent and persistent until he or she has addressed all of your concerns to your satisfaction.

I know that in this day of demanding workplace environments and rising health-care costs (including insurance premiums, co-pays, and office visit costs), you may say you don't have the time or money to spend at the doctor's office. I say this is an investment in your health and you don't have the time or money not to. If you have followed step one, you have already found a doctor with whom you feel comfortable. You must think of this as any other important relationship and invest in it for the long term.

Prescription medications

It is essential that you know detailed information about your prescription drugs. You must know the medication name, generic name, strength or dosage, and how often you must take it. If you cannot remember, you must have the name legibly written so anyone can read and understand it. Not only should you know the name and dosage of each medication, you should also know exactly why you are taking it. Do not take anything that you do not understand. At the time your physician writes the prescription, you should understand the purpose of the medication, the duration (how long you have to take it), and its common side effects.

Tests

Also know why you are taking any tests that are ordered, including blood tests, and the side effects of any test procedure. For example, any procedure using IV dye or contrast can worsen renal (kidney) function. Once you have completed a test, follow up on the test results, and discuss with your doctor when to expect them. If you do not receive your results during this agreed-upon time, you are to seek them out yourself, usually from your primary care physician but sometimes from the testing center itself. More than one physician has not relayed test results at a promised time for a variety of reasons: they were lost in the mail, office staff accidentally discarded the paper copy, or the testing facility never sent the results, to name a few.

Communicating with your doctor

Use alternative methods to contact your doctor. Telephone calls can sometimes be time-consuming and cumbersome for a busy physician. Let the physician know if he or she can leave a voice mail or answering machine message if he or she is unable to reach you. Also, consider consulting with your doctor by e-mail. This may be the most convenient for both of you. Some doctors even offer an electronic office visit, where patients can pay a small co-pay for an acute consultation through an e-mail-type system.

Finally, if you have a family member who is illiterate or has difficulty reading and understanding information, you must take extra steps to ensure that he or she receives the best possible care. Make every effort to attend each office visit with your loved one. Some doctors may even consent to recording the plan for the visit on a small recording device so the patient and any other necessary family members can play it back. Please do not tape-record any visits without permission from the physician and the patient involved. Both must to be willing participants in this kind of interaction.

Respect This Relationship and Make It Important to You

Please remember to be on time, preferably ten to fifteen minutes early. You may just be able to be seen early, but, chances are, your doctor will be running late, so plan for it. We cannot predict medical emergencies, which may occur daily in a busy practice. If you are not seen within a timely fashion, it is okay to ask if you should reschedule or if you have been forgotten or possibly misplaced. Sometimes, charts, lab work, and even patients get misplaced. If you are unsure why you are not being seen on time, please ask, but, more than likely, your doctor is probably just running behind.

Attitude

Your doctor can either be your best advocate for your health care or your worst obstacle. If every time you come in for an office visit you have a sour look on your face and your approach is confrontational, your doctor or the other office staff and health-care professionals may not enthusiastically welcome your visit. You will become "that patient", and the goal will be to complete your visit as soon as possible and expedite your departure from the clinic. You want your visit to be a welcome addition to your doctor's day. All physicians have some patients that they cannot wait to see. You want to be one of those.

Be pleasant to the other medical and office staff. They have more power than you realize. The front desk people can get you on a doctor's packed schedule. They have the ability to find an appointment when it appears that none is available. It is to your advantage to treat these people with respect and dignity. The respectful patients are the most accommodated and seem to get in for even specialist appointments and referrals when no one else could get them in. Remember, these front desk people usually have been working in their current capacity for quite some time and know a lot more about the system (HMOs, hospitals, clinics, ERs, specialists, and so forth) than even many of the doctors. They are often aware of back doors. They may know people in different departments who are able to make sure you are seen quickly. Do not allow a negative attitude (yours or the front desk person's) to contaminate this relationship.

Yes, you are the patient and the priority in this health-care relationship. No health-care worker will intentionally mistreat you, but everyone is human. Remember the old saying, "You can catch more flies with honey." You don't want anything you can control to become an impediment to you receiving the professional medical care you deserve.

If you do have an unpleasant experience or anyone in the health-care office (for example, doctor, nurse, medical assistant, student, front desk person, or person on the telephone) mistreats you, you must speak with a supervisor, nursing manager, clinic manager, or other person in charge. Mistreatment by anyone in the health-care field is unacceptable at any level. You are a customer, and you have the freedom to choose who will provide you the service of caring for your health. It is important that you are treated with dignity and respect, but, if you start with a positive attitude of your own, you will get much more than you expected.

Last, but certainly not least, please avoid ending the visit with, "Oh, by the way ..." There is little that gets under a physician's skin more. This generally occurs near the end of the visit, when the doctor is wrapping up the appointment or may even literally have a hand on the doorknob to leave. This is when the patient will state something tremendously important or embarrassing, and it is usually the real reason the patient has come in to be seen. It may be a concern of a sexual nature, such as a sexually transmitted disease (STD) or erectile dysfunction—something the patient may be very embarrassed by, which is why they might have put off mentioning it.

Believe me, most physicians have heard it all before. You are probably not the first patient your doctor has ever seen with your concern. Please be up-front with all of your concerns so they may be addressed initially and completely during the visit. If you cannot bear to say it, then record it, write it down, or type it out.

Ask for What You Want and Be Direct

There is a fine art to asking for what you want. Please do not demand things from your doctor. When any person feels another person is pressuring him or her, the natural tendency is to push back and not succumb to the wishes of others or give in to the aggressor. Doctors feel the same way. It is important to be direct but not pushy and overly aggressive.

It is never a good idea to threaten any physician or other office worker with legal action. This may temporarily get you what you think you want, but, in the long run, the staff will begin to avoid you, and you will develop an unhealthy relationship with your physician. You don't know what exactly you need. If you did, you wouldn't need to come in to be seen. Just because a certain treatment or medication works for your co-worker, friend, or even a family member, it does not mean it is the correct treatment for you. Please trust your doctor. If you do not trust him or her, find another doctor. If you threaten your physician, you break the trust, and your doctor will probably always be uncomfortable with treating you thereafter. In the end, your doctor may relent to your requests, which may cost you money, worry, and possible negative side effects, including unnecessary medications, testing, and procedures.

Please bring in all of your research and questions. As a group, doctors welcome intellectual challenges. It is part of the reason most of us entered the medical and profession. We can discuss whatever you wish, but being rude and disrespectful is never appropriate. Asking until you have sufficient understanding is always appropriate. You must find a physician you feel comfortable asking questions. When your doctor doesn't know the answers, It is important that he or she is willing to tell you that and be prepared to find the answers, if they exist. Some questions either don't have any answer or have multiple answers. To other questions, the answer may be a resounding "no", this can be a valid and appropriate response as well. You want a physician you can trust to tell you the truth and not just what you want to hear.

Believe

If the doctor diagnoses you with a medical condition, get treatment. If you do not believe the diagnosis, get a second opinion. Once you get a second or even a third opinion and the answer is the same, please begin treatment. If the diagnosis depends on something objective such as numeric values (for example, diabetes, hypertension, or hypercholesterolemia), accept it and get treatment. If the diagnosis depends on clinical judgment and other criteria (for example, lupus, chronic fatigue syndrome, or fibromyalgia), It is okay to question the diagnosis and seek another medical opinion before treatment. Treatment may mean taking medication, completely changing eating habits, exercising regularly, lifestyle counseling, therapy, or any combination of the previous. Accept your diagnosis and begin fighting the condition, not your doctor.

Pursue, Pursue, and Pursue

If something does not seem right to you, pursue it. If you have not heard anything about a recent test, exam, or lab work, please call or visit your doctor's office for the results. No news is not always good news. Lab results do get misplaced, incorrect prescriptions are written and called in, and errors do occur. If you feel something is not going well or is just plain wrong, speak up. Please discuss the concern with your doctor. Before you leave the examination room or office, ask when you should expect to receive your test results or when to follow up if you are not feeling better. As the patient, you must take your own health care into your hands. If you do not receive your results when you were told to expect them, then it is time for you to pursue them on your own. It is okay to ask. The test results are yours. Asking for something that you did not receive at an agreed-upon time is not being pushy or overbearing. It's being responsible. You may ask for your results by telephone or visiting the office to pick them up, but do not expect to speak to your doctor without an appointment.

Set Realistic Expectations

Your doctor is not a miracle worker. You must do your part and take the advice given. If you have not taken the responsibility of your health seriously for the past twenty years, you and your doctor may have a lot of work to do. You will not be able to undo twenty years of damage in one office visit. It will take time. If it took years to put on the one hundred pounds you've gained, it won't come off, at least in a healthy way, in one day, one week, one month, or even one year. It is a gradual process. You can't expect one person to undo years of abuse you have inflicted on your own body.

Be Open-minded

Your doctor may offer suggestions for your treatment that may not be what you had in mind. Take the information, do some research, and get a second opinion, if necessary. If you don't have all of the answers, don't discount the treatment regimen at first glance.

Your doctor may also practice in a teaching hospital. This means that the doctors and medical students he or she is teaching may be assisting him or her. These doctors being taught are residents, interns (first-year residents), or medical students. Medical students are truly students. They have completed a college undergraduate degree and are now in school to become medical doctors. Residents and interns are already medical doctors. They have all completed medical school successfully. Seeing a resident or intern is not something you should consider negative. On the contrary, consider this a positive experience. You will have at least two physicians reviewing your medical case. More than once, an astute resident has brought a new treatment or diagnosis to my attention. Also, residents' schedules are usually less dense than those of senior staff physicians. This means you will usually have more time to spend with your doctor. Finally, residents rotate through multiple clinics and departments and get information as it is discovered. They rotate through disciplines such as cardiology (study of the heart), gastroenterology (study of the esophagus, stomach, and colon), ENT (study of the ear, nose, and throat), and infectious disease. Each time they rotate, the residents carry away the cutting-edge procedures, treatments, and methods of diagnosis. Residents and medical students keep your doctor fresh and current. We all have to stay current in order to teach. It is to your benefit to see the resident or the medical student along with the senior staff doctor.

No medical student or resident can ever legally practice without supervision. Your case must be specifically reviewed and assessed with the treatment plan that the senior staff physician presents, manages, and approves. Your medical condition and chart information will be reviewed in detail as if you were primarily seeing the senior staff physician.

Please be patient with these medical students and new doctors. It may require more time for your office visit, but they usually have time to uncover information—family history, social history, and other concerns—of which your doctor was never aware. You are getting more than you know for your one office visit. Also, remember your doctor was once a medical student, an intern, and then resident. Now you have a great doctor partially because of this professional maturation process. So relax and enjoy all of the attention.

Uninsured and Underinsured

If you are working but not covered, seek individual health-care insurance coverage. Compare individual policies on the Internet, along with their costs and benefits.

If you are close to losing or have recently lost your medical insurance coverage, tell your doctor. He or she:

- Will be able to write prescriptions for all medications you use on a daily basis (maybe even up to a one-year supply)
- Will have recommendations for current and future care
- May even be able to continue to see you for a reduced price

If you are not working or cannot afford individual health insurance, it is important that you seek assistance from free or reduced-price clinics. You may begin at your local health department for further information. Look for assistance from local churches and charities. Call around and attempt to make an appointment. You may have to wait a month to get one, but, when you get an appointment, show up. Be prepared. There may be long waits and long lines. You may not be seen the first time you go to the clinic, but, without insurance, it is even more important that a doctor see you. You require health maintenance and prevention. If you have an urgent concern, state it, and don't be shy. The receptionist may be able get you in to see the doctor quickly. Be pleasant, patient and persistent. It may take more than one or two visits to the clinic to be seen.

No matter how long it takes for you for you to get an appointment, you should always be treated with respect and dignity, even if you don't have insurance. If you are not treated respectfully, you can air your concerns with a supervisor. Even if you are seeing a solo practitioner, you can always speak to someone about inappropriate behavior. Utilize and speak with supervisors if necessary, even if you are not paying for your care. Lack of financial means does not mean you are a second-class citizen. You deserve complete and adequate health care.

If you are paying with cash, discuss this with your doctor. It is important you receive the least-expensive but effective medications, and you may need to have only very specific lab testing to get more for your money. Your physician may be able to save you a significant amount, and some clinics offer a discount, usually 20 percent or more, for patients paying with cash.

If prescriptions are more expensive than you can afford, tell the doctor, and do not be embarrassed. Sometimes, physicians don't know the cost of the medications prescribed. We just focus on treating the disease process. Your doctor won't know if you don't speak up. Your doctor can give you a prescription for generic or lower-price medications to assist with cost. You can also seek assistance from discount warehouse clubs, grocery stores, or other discount stores.

A brief note here regarding generic medications: Generic medications have the same active ingredients as brand-name medications. Utilize this fact, and ask for a generic equivalent for any medication. Also, remember that new medications are set to become generic each month, so keep checking with the doctor and the pharmacy. They will often know when medications are likely to become generic. Check each time you get your prescriptions filled. You may be pleasantly surprised. You are not being a nuisance; you are being cost-conscious. If you don't ask, you may spend your hard-earned money unnecessarily.

Avoid using emergency rooms for nonurgent common health concerns such as coughs, colds, ear pain, vaginitis, or bladder infections. Your local ER will charge you as much as ten times more than a visit to a doctor's office or a local clinic. The emergency room is for exactly that, emergencies only. If you truly believe you have an emergency, then absolutely be seen immediately. If not, go to your local primary care clinic.

* * *

This is where you start. You can and should approach your relationship with your doctor in this way. Use this information for yourself, and pass it on to friends and family.

Part Two

Let's Go!

Chapter One

Mrs. Collins

Today, I will see twenty-six patients, thirteen in the morning and thirteen in the afternoon. That may not seem like a lot, but it is when you also have eight telephone messages to return, prescription orders to refill, thirty to forty labs, x-rays, stress tests, and abnormal Pap smears to interpret, and other day-to-day medical duties to follow up. I work in the section of an urban city that is slowly making a comeback, but has definitely seen much better and more prosperous days. The city has unfortunate and saddening poverty. People are robbing Peter to pay Paul, as the proverb says—but Peter doesn't have any money either.

I receive a telephone message stating that Mrs. Collins' blood sugar is high and she needs to be added to my already-full schedule. Mrs. Collins, a seventy-six-year-old female, has been diabetic for about ten years. She takes the maximum doses of her oral medications, and we have been discussing insulin for the past four years. Mrs. Collins has been adamant that she will not give herself insulin injections, stating her mother and brother lost their feet to amputation after they "got on that needle." We've had multiple conversations about diabetes and the nature of the disease. She understands her relatives probably were as resistant as she is now to start insulin, and they both started too late. The vascular damage had already been done, allowing little to no blood flow to these extremities.

Mrs. Collins arrives about twenty minutes later. After we exchange our brief hellos and she is ushered into an examination room, Mrs. Collins begins a lengthy discussion about why her blood sugars were elevated. She apparently had begun her day as usual, awakening at six in the morning. She ate a breakfast of oatmeal, toast, and an apple and then realized she was low on groceries and needed to go to the grocery store. She states it was a good thing her pension check would be coming soon because she was low on cash and much too proud to approach her daughter for money.

Mrs. Collins is unique. She's one of the few of my elderly patients with family members who actually care. She comes from what appears to be a very loving family, some of whom I have met. Many of the elderly patients I treat do not have anyone.

Mrs. Collins is also part of a growing population of adult senior poor who have worked their entire lives and now find themselves not having enough for day-to-day expenses. She had about $15 for grocery shopping and bought inexpensive things that she could afford, which were all high in fat, sugar, and salt.

After speaking with Mrs. Collins for quite some time, I begin to understand that she didn't mean to be noncompliant. She just couldn't afford the healthy alternatives of soy supplements and tofu that the nice dietician here in the clinic offered to her.

After I evaluate her blood sugar, which was, in fact, high, Mrs. Collins does receive a small dose of insulin in the clinic that day. I once again explain my concern for her heart, kidneys, brain, and extremities. We talk about some simple changes she could make at the grocery store such as buying ripened (but not rotten) fruit for a reduced price, clipping coupons, and shopping at some of the discount stores and warehouse clubs. Because many of my patients are senior citizens, I find myself increasingly worrying about what they eat. If I'd had more time,

I would have gone to the grocery store with Mrs. Collins to help her pick out inexpensive healthy alternatives to the food she'd been choosing. She thanks me and leaves the clinic.

I believe the most important thing I did for Mrs. Collins that day was to call a family member after receiving her permission. I explained the situation to her daughter, who was quite concerned, yet knew nothing of her mother's advanced diabetes or inability to afford appropriate food. The telephone conversation was important because, in addition to financial assistance, Mrs. Collins' daughter was able to commit to helping her mother with weekly grocery shopping and making sure she had what she needed. The extra fifteen minutes it took for me to speak with Mrs. Collins' daughter quite possibly may have saved her a future hospitalization for uncontrolled diabetes.

Diabetes

I can thank Mrs. Collins for allowing me the opportunity to review and discuss multiple facts about diabetes.[1] She is an excellent example of what is occurring in this country. Diabetes, a nasty, progressive, crippling disease, claims the lives of millions, both directly and indirectly. It doesn't have to be this way.

Diabetes is primarily a vascular disease that occurs in two ways:

- The first is **type 1**, which generally occurs in children, when the body mounts an immune response and literally attacks the pancreas (the organ that produces insulin).

---

[1] And poverty, retirement, Medicare, and improper treatment of seniors, but I will limit this discussion to diabetes.

- The second and more common is **type 2**, which occurs when the body will not utilize the insulin that the pancreas already produces, usually in an overweight person. This patient is insulin resistant.

Your natural insulin keeps blood sugar regulated both with and without food intake. When the body becomes resistant to insulin, it will not respond to natural insulin, and blood sugars rise. When blood sugars rise and remain elevated, these sugars become toxic and cause damage to all of the blood vessels in the body. This is the reason for the multiple negative effects of diabetes, including damage to nerves in eyes, kidneys, heart, and extremities (legs, feet, and hands). This is why:

The leading cause of preventable blindness in the United States is diabetes.

The top diagnosis of patients on dialysis is diabetes and hypertension. (We will talk about hypertension later.)

Diabetics have a risk of heart attack equal to that of a person who has already had a heart attack. (Yes, it is that serious.)

Diagnosis

An accurate diagnosis is important. A diagnosis of diabetes now requires one of the following:

- Two separate fasting blood sugars of 126 or greater
- A blood sugar greater than or equal to two hundred at any time with symptoms such as polyuria (urinating a lot), polydipsia (drinking a lot), or polyphagia (eating a lot)
- A positive two-hour glucose tolerance test, or
- Two elevated hemoglobin A1C tests

When you are first diagnosed, you may need to see your doctor at least once a week until your blood sugars are controlled. You will need a blood test (a hemoglobin A1C) as often as every three months, but, minimally, every six months. This test measures blood sugars as an average over a three-month period. When blood sugars are high, the glucose molecules actually attach to red blood cells. The higher your blood sugars, the more glucose will be attached to the red blood cells. It is possible to determine what percentage of red blood cells are glycosylated (covered with glucose). This number, the hemoglobin A1C, is expressed as a percentage of all red blood cells. In diabetics, this number should be below 7 percent. Normal is about 5 percent.

If you are diagnosed, the numbers don't lie. You can get a second or third opinion if you like, but, if it is the same, accept it, and believe it. Actually accepting, believing, and taking control of your illness is probably the first step toward treatment. Being diabetic is not the end of the world, but not taking care of yourself can make diabetes the end of YOUR world. If you need professional counseling, get it. If you need to talk to someone, talk. You really have no excuse for not taking care of yourself.

Managing diabetes

It is important that diabetic patients, along with their family and friends, know how to manage the disease. In this book, I repeatedly mention the benefits of changing the food intake or diet and the benefits of exercise. I use the word "diet" in reference to the foods you eat. If most of us could get these two things correct, we could prevent, or at least cut in half, most of the common chronic diseases.

Caring for yourself or a loved one with diabetes requires discipline, control, and commitment. This is change will occur overnight. You are in this for the long haul. You will have some ups and downs, but, overall, most people can control their diabetes.

If you have diabetes, you must get serious about your disease. Only you can prevent the complications. You must exercise, eat better, take your medicine, and start insulin, if necessary. Insulin does not cause people to go on dialysis or to lose their limbs, but uncontrolled diabetes does. We should think of insulin as our friend, not as the punishment for a life of bad behavior. You may not have done anything wrong. Sometimes, even when people do everything right, including consuming proper foods, and getting exercise, they still become diabetic. Diabetes is a nasty disease that will often progress on its own, even if you are doing everything right.

After you accept the diagnosis of diabetes, you must (after receiving your doctor's approval) be able to commit to a schedule of regular exercise. Optimally, you should get twenty to thirty minutes (at a time) of aerobic exercise five days a week. A leisurely walk to the corner park is excellent for blood flow and overall well-being. It relieves stiffness and assists with other ailments, but that is not aerobic exercise. Aerobic exercise is any activity that raises the heart rate including activities like fast paced walking, running, biking, swimming and aerobics classes to name a few. This is an exercise intensity where you are able to speak in only short complete sentences and may only speak in short phrases (such as yes, no, or "I don't know"). If you are able to have a phone conversation during your exercise, you need to be at a higher intensity.

Next, you need to change the way you eat, not start a diet. The word "diet" implies a temporary change in food intake to lose weight, then return to previous eating habits. No, you must change the way you eat, drink, and think about food. You should begin with a dietician or other professional to help you make better choices. When you see this person, you must discuss your living situation along with your budget, career/job responsibilities, household responsibilities (that is, marriage, children, and primary meal preparer), family (any family members with diabetes), and friends (anyone with diabetes or giving you opinions about what is good and not so good to eat).

There is not a team of gremlin doctors around whose only mission is to get you on insulin. I wish more people understood that most physicians do not work for pharmaceutical (drug) companies. We do not get any incentive or benefit for putting you on the medication you need. If in doubt, just ask your doctor if he or she:

- Has a relationship with any pharmaceutical company
- Is being compensated in any way for prescribing the way he or she does
- Receives free lunches, expensive gifts, or other perks

You are entitled to know the answers to these questions. A yes answer may not make a doctor someone you can't trust. Your physician may receive compensation simply for prescribing a drug he or she would already be using extensively in his or her practice. It is important that you know this information up front. You should be able to choose your primary care physician with all necessary information fully disclosed. Doctors are obligated to tell you the truth. All you have to do is ask.

If you are on prescribed medication, you need to take it. Believe it or not, you are not the best judge of whether you need medication. Your doctor is. Even if you are able to change your diet and exercise and get your blood sugars down, you may still need medication. You will also need to get a glucometer (a machine that measures your blood sugar) and use it as often as your doctor requires. Yes, poking your fingertips is uncomfortable, but is necessary. You can only accurately manage your medications with consistent measurement of your blood sugars.

If you are to take your medication twice a day, then take it twice a day. If you honestly feel that you can only remember once a day, then tell your doctor, who may be able to coordinate a regimen requiring only once-a-day dosing. Trust your doctor enough to know that he or she has your best interest at heart. Remember, you took care and time to choose this person.

Cholesterol

Diabetic patients should have a fasting cholesterol test at least once a year and more for uncontrolled cholesterol. This is not so much for diabetes treatment as it is to manage an important risk factor for heart disease. The fasting cholesterol test should include measurement of your low-density lipoprotein (LDL), which is most closely associated with coronary artery disease or heart attack. This number should ideally be below seventy, but definitely below one hundred. You may need cholesterol-lowering medication to achieve this level. Changing the foods you eat and exercising may not be enough, so, if you need the medicine, take it.

Many people are concerned about the side effects of these medications, so, by all means, discuss your concerns with your doctor. Cholesterol-lowering medications can cause liver damage (which is reversible upon discontinuing the medication), muscle discomfort, and other side effects. Being a diabetic causes you to have a high risk of heart disease. To minimize your risk, you may need cholesterol-lowering medication. As a patient, you must put your fears in proper perspective. You must be more concerned about having heart disease or a heart attack than the possible side effects of medication that may or may not occur.

Kidney disease

Diabetic patients will require a yearly test to check for protein in the urine, an early marker for kidney disease. To attempt to prevent kidney failure and the progression to dialysis, a patient may be placed on medication for blood pressure, which must be kept at lower levels in diabetics than hypertensive patients in the general population. The National Kidney Foundation recommends that blood pressure be kept below 130/80. Some studies are now suggesting even lower blood pressures (below 125/75) are necessary for non elderly African American persons, to best prevent kidney disease. Better blood pressure control has been shown to slow the progression of kidney disease and to decrease the need for dialysis. I have a full chapter on hypertension later in this book. Diabetics may be placed on medications called ACE inhibitors or angiotensin receptor blockers (ARBs) that offer both blood pressure control and kidney protection by decreasing proteinuria (proteins escaping into the urine from the kidney). Lower dosages of these medications may be necessary in diabetic patients to protect the kidneys even if blood pressure is well controlled.

Foot care

At least once yearly, diabetics will require formal foot screening in the physician's office, but basic foot care and screening is primarily done at home. A patient with diabetes is never to walk around barefoot. Diabetics must check their feet daily for any cuts or sores. The patient must consult with the doctor about whether it is safe to cut his or her own toenails and remove foot calluses. The annual screening is done in the primary care doctor's office with an instrument called a microfilament, a thin, wirelike device used to check different areas on the foot for decreased or lack of sensation. The doctor will also check for appropriate circulation in the foot as well. It is important that a diabetic patient take foot care seriously to prevent infection and ulceration, which can lead to amputation.

Eye care

Eye exams are required at least once annually and possibly more often, depending on an eye care professional's recommendations. An evaluation performed by a person trained in the examination and treatment of diabetic eye disease is essential. A dilated retinal exam is necessary. A simple exam for glasses is not enough. This special eye exam allows the doctor to look into the back of the eye to check for any areas of concern. It is possible to detect and correct some diabetic damage, if detected early, before damage has progressed.

Summary

Fully managing and treating diabetes requires the coordination of all of the above. With dedication and commitment, it is possible to live a healthy and fulfilling life with the diagnosis of diabetes. Your health-care team is ready and waiting to assist you! You are the important factor here in prevention and treatment. Speak with your doctor, ask the questions, and get the answers.

Chapter Two

Mrs. Johnson

It is now eleven in the morning. I have a one-thirty meeting, and I'm running late as usual. My ten-twenty patient, who has lost all patience, wants to leave or reschedule. After finishing with my ten o'clock patient at ten forty-five, I finally turn my attention to my ten-twenty patient. I'm running at this point. My adrenaline flows; my mind races. I rush into an exam room to find an exasperated Mrs. Johnson sitting and reading a two-year-old copy of *Reader's Digest.* (I make a mental note to get more current reading material.) I observe the examination table and see the crumpled exam table paper in the spot where Mrs. Johnson probably sat waiting for at least thirty minutes. I apologize sincerely to Mrs. Johnson.

Through pressed lips with a hissing quality to her voice, she responds, "Yes, I have been waiting quite a while. I almost left."

I'm eager to change the subject. "How are you doing? What brings you in today?"

After a brief discussion about her granddaughter's recent graduation and her husband (also a patient of mine) and his constipation, she begins to relate a few of her spiritual insights and beliefs, which require her to refuse many treatments offered in Western medicine. Mrs. Johnson has markedly uncontrolled hypertension (high blood pressure), yet refuses to take any antihypertensive medications. Instead, she prays about her blood pressure and discusses it with her church members during group prayer. Mrs. Johnson does not want to discuss medication options, and I, once again, find myself sinking into a preachy conversation in which I extol the virtues of taking medications and the dangers of not doing so. Mrs. Johnson refutes every point and continues her strong conviction against medication, stating she will return for further evaluation and an additional blood pressure check because she is sure that her blood pressure is up because of "that pork" she ate earlier.

Mrs. Johnson is unknowingly waiting for a stroke, heart disease, vascular disease, and/or kidney failure to happen. I'm genuinely afraid for her and tell her so. I could not bear to have Mrs. Johnson on my conscience, so I sent a home health-care nurse to her residence, who offered her diet counseling for sodium restriction with the hope of lowering her blood pressure without using those "toxins," as she called them.

Although Mrs. Johnson tried hard to control her blood pressure on her own, she was unable to tame her hypertension and did suffer a debilitating stroke. When I saw her during the first follow-up visit after being hospitalized, she was a sad shell of a woman with a droop on the right side of her face. Her speech was noticeably slurred but intelligible. In any other situation, an "I told you so" might have been in order, but not this one. I wanted to cry, scream, and kick all at the same time. I had told her this might happen, but wished she would prove me wrong.

Hypertension

Hypertension is defined as a systolic blood pressure (the top number) of greater than 140 or a diastolic (the bottom number) of greater than 90. Pre-hypertension is blood pressure between 130/80 and 140/90. Each patient must discuss with his or her doctor what a safe and healthy blood pressure is for him or her.

Control of blood pressure is vital to overall body function. The brain, heart, kidneys, extremities, and eyes are being damaged, possibly beyond repair, every single day that Blood pressure is not controlled. Yes, this may also be the reason for erectile dysfunction. Hypertension is more important than you believe.

Hypertension is dangerous, and, if left untreated, may lead to heart disease, kidney failure, or stroke. When people don't feel bad, they don't see the need to seek treatment. Because hypertension is the silent killer, patients frequently don't feel bad until it is too late—when they are unable to move or speak express thoughts or when the speech becomes an unintelligible salad of sounds. This is not something anyone wishes to experience ever, and, although strokes are devastating, prevention can be possible.

Managing hypertension

In some cases, you can control hypertension with aerobic exercise and a change in diet. If a response to salt triggers your high blood pressure, simply cutting your salt intake to a total of 2,000 milligrams or less a day may help control it.

It is possible that vitamins and herbal supplements may be valuable assistive agents in Controlling blood pressure and other chronic conditions, but they also can cause side effects or interact with other medications. Please discuss with the doctor any and all over-the-counter medications and supplements you are taking. This information is vital. If the vitamins or herbs are safe, the doctor will not prohibit them, and most doctors will actually support this. Physicians are not trying to push pills; there is not a conspiracy to get you on medication. But supplements may not be enough to completely control your blood pressure, and prescribed medication may be key. Taking prescribed medication is necessary if your blood pressure is not controlled after an adequate trial of exercise, weight loss, diet changes, and/or supplements. Getting multiple blood pressure readings is appropriate; however, if your blood pressure is dangerously high, you may need medication immediately, that is, at the first visit. With early treatment, it is possible to prevent many complications from this disease.

It is impossible to become addicted to blood pressure medication. Although it is rare for people to be able to discontinue their medication, it does happen periodically. When patients are serious about their diet, exercise, and weight changes, I have seen medication requirements lowered and sometimes even discontinued when those changes were enough to control blood pressure. This is rare. More often, one blood pressure medication is not sufficient, and a second, third, or even a fourth medication may be required.

Finally, many people feel they will control their blood pressure by praying about it only. Praying about it may not be enough. I'm not discounting the power of prayer; I'm simply saying that spirituality is an individual gift to each person. You should not view taking medicine as a failure of your spiritual faith. It may be just the opposite: it could quite possibly be your spiritual faith warning you and letting you know it is time to be compliant with your treatment regimen, whatever it may be. Keep in mind that there is a reason that doctors have come into your life. You may be at this moment in your life for a time such as this, to meet a doctor such as this.

Summary

Preservation of your life and functioning are more important than anything else. First and foremost, take care of yourself. It may be the smartest thing you have ever done. If anyone you know is displaying signs of a stroke, such as difficulty speaking, facial or body weaknesses, or dizziness, please call emergency services immediately. Emergency services now may help with retention of function in the future. You are the important factor here in prevention and treatment. Speak with your doctor, ask the questions, and get the answers.

Chapter Three

Justin

"Doc, I need you to look at somethin', I got a bump", greets me as I enter the exam room.

Justin is a young man in his early twenties who asks me to call him by first name only. I sit on my office stool and prepare to discuss the possibilities of this still-unseen lesion, along with medical history and circumstances surrounding the appearance of the "bump". But there isn't much time for discussion because he quickly loosens his belt, abruptly drops his pants, and Stands in front of me while diligently searching his genitals for this "bump". He wants immediate reassurance and understanding about just what this is.

I ask my patient to lie on the examination table, and I quickly sanitize my hands. I put on a pair of gloves and examine the area in question. I discover a small lesion on the side of his penis. I have grown fairly used to evaluating and examining this lesion, a condyloma (genital wart), over the years. I don't want to tell this twenty-five-year-old man that he has genital warts. We discuss the possibilities of the lesion. I decide to take a biopsy of it and send the sample to the pathology lab for evaluation.

Of course, the pathology is positive for a genital wart, and now we must have the conversation. Let it be known that doctors do not like to have these conversations. The sexually transmitted disease (STD) talk is not fun. It is especially not fun when there is no real cure, only a treatment, for the disease.

I call the patient and discuss his options. I have already prepared him with the possibilities during his previous clinic visit. He thanks me for the call and the education. I call in a prescription and ask him to come back for a follow-up appointment in one month. I haven't seen him since. He hasn't answered my calls and voice mails.

STDs

Let's talk about sex. Whoa! Where did that come from? Well, patients need to know the truth about sex and STDs. Unless you and your partner are each other's first everything (including kissing, oral sex, and genital sex), you both have probably been exposed to one or more STDs, even if you don't know it. Even if you have been using condoms. Whoa! Does this mean that STDs can be transmitted even if you or your partner wears a condom? My answer to you is, "Yes, absolutely."

You can catch some STDs simply from skin-to-skin contact (the parts that a condom doesn't cover). Herpes and genital warts are two such diseases; others generally require some contact with vaginal fluid or semen. Whatever the infection, it is not something to which you want to be exposed. There are medical and spiritual reasons that we should all wait until marriage for sex, but let's be honest. Most of us don't.

If you are going to engage in sexual intercourse, you must use condoms every single time. Condoms greatly lessen your chances for catching a STD. Notice, I said "greatly lessen," not "prevent." Nothing except abstinence can prevent STDs. We must keep ourselves as safe as possible. It is important to be tested for all of the common STDs, including gonorrhea, chlamydia, trichomonas, hepatitis B and C, syphilis, herpes, and HIV. You can even catch cervical cancer from an STD, but I will discuss more about that later.

Talk to your doctor about sex, and be honest about your experiences with diseases, sexual partners, and abuse (domestic violence, physical, and/or sexual) of any kind. This will not shock your doctor. Unfortunately, you are not the first person to go through the situation you are in. We have many resources that may not be available for you except through the doctor's office. Be honest with the doctor. We can only help if you let us.

Remember, your conversation with your doctor is between the two of you only and, therefore, confidential. You must give written consent for your medical history to be shared, with the only exception being a life-threatening emergency or situations involving abuse. Legal policies require criminal prosecution for any breech of information outside of the health-care team. Not only can employees be terminated, they can face hundreds of thousands of dollars of penalties as well as jail time for any misuse of your medical information. Your information is safe. It is used only to help, not to judge or persecute.

Summary

Both men and women require screening. If you do have symptoms (and many people will not), they might include fever, vaginal discharge, penile discharge, burning with urination, or abdominal pain. Your doctor must check you for STDs. It is possible to be infected for many years and never have any symptoms. If you are going to have sex, take the responsibility and get checked. Begin the discussion with your doctor, follow through, and don't be embarrassed. Embarrassment, ignorance, and ego keep infections running rampantly through the community. This is not necessary. Let's take care of each other. You don't even have to have the "correct words." If you can't bear to say it, write it down and give it to your doctor. Find a way to communicate it.

Chapter Four

Ms. Jakes

"What is an abnormal Pap smear? Do I have cancer? What is a colposcopy?" Ms. Jakes asks as she sits on an examination table.

She is completely bewildered with a fearful expression on her face and guarded body stance. Ms. Jakes definitely had a difficult life by anyone's standards. A close family member sexually abused her, and she had two children out of wedlock with a very abusive man. She has worked at one of the auto plants for ten years and makes a more-than-decent living. She has finally found happiness in a monogamous same-sex relationship with a woman she met at her job, and things seem to be going well for her. She continues to come to the office only because of the relationship we have built over the four years I have been seeing her as a patient. She has never felt she needed Pap smears because she hadn't had sex with a man in more than five years.

After much coaxing, cajoling, bribing, begging, and pleading, Ms. Jakes had a Pap smear two weeks before her office visit, and she is now coming in for the test results. She agreed to have the test only after I promised her that, if it were normal, I would not bother her about another Pap smear for quite some time, that is, another year. I was concerned because Ms. Jakes had not had a Pap smear since her seven-year-old son was born. She gave up sexual activity with men after ending her relationship with her son's father five years ago. A year later, she met her current partner, and they have been together and raising her two children for nearly four years.

"Do I have cancer?"

"No, Ms. Jakes, I'm not saying you have cervical cancer, but I'm saying your Pap smear was not normal and you have some early, precancerous cells that could turn into cancer if we do not treat them. You need to have a test called a colposcopy. This is a test where we look at your cervix under a microscope and take pieces of your cervix that do not look healthy. Those small pieces, or biopsies, are sent to the pathology laboratory to determine if they have the potential to turn into or are early cancer."

"How did I get this?"

"Well, we know a virus causes most cervical cancer. Just like you catch a cold or the chicken pox, a virus also causes this. It's called the human papillomavirus (HPV). Not only does it cause cervical cancer, it causes genital warts as well."

"I never had warts," Ms. Jakes declares.

"I know you haven't, but you have contracted HPV through sexual intercourse. This virus is a very sneaky one and requires skin-to-skin contact, which can be transmitted despite the use of a condom."

"Oh, so can my partner catch this?" asks Ms. Jakes.

"Yes, it is possible."

"So now what?"

Exactly. Now what?

Cervical Cancer

Located deep inside the vagina, the cervix, the entrance to the uterus, is a living, breathing organ. It changes from birth to puberty to pregnancy to menopause. After exposure to HPV, changes occur in the cervix causing the cells (very small areas) on the cervix to change and become cancer cells. All cancer cells are normal cells that grow excessively and out of control. Many times, the body can fight HPV, and no long-term damage occurs, but the virus sometimes has the optimal environment to cause cell damage, and it transforms normal cells into cancerous ones. If medical care is received early in the process, we can eradicate (cut, burn, or freeze) or destroy the abnormal cells. In this way, it is possible to prevent cervical cancer.

This is why a Pap smear, the gentle scraping of the cervix from the portion where the cells change or transition rapidly, is so important. This transition zone is where most cancers of the cervix originate. If the Pap smears are normal for three consecutive years, this is a good indication that there is probably no evidence of cervical cancer, but this is not absolutely true. It is possible to have very slow-growing cancerous cells, which develop over years with the same sexual partner. You will need to discuss your optimal timing for cervical cancer screening with your doctor.

For women who have had a hysterectomy for cancerous reasons, cancer of the cervix, uterus, or ovaries, an annual Pap smear may still be recommended. For women who have had hysterectomy for benign or noncancerous reasons, including fibroid tumors or pelvic pain, there is generally no need for further Pap smears. You must also know if you had a supracervical (above the cervix) hysterectomy, the removal of the uterus while leaving the cervix intact. In this case, regular annual Pap smears are still recommended. It is your duty to get an understanding of what your timetable will be for getting Pap smears and the reasons why.

Once a Pap smear is taken, the sample is sent to a pathology lab and read under a microscope to determine if the sample appears normal. In some labs, an actual person reviews the sample. In others, a computer completes the review. A pathologist (a medical doctor trained in microscope detection of disease) generally reviews the abnormal samples. The pathologist will give an educated opinion as to whether either precancerous or cancerous cells are present.

If it is determined that your Pap smear is abnormal, you should have a test called a colposcopy, an examination of the cervix under a microscope with special lighting and stains to determine whether abnormal cells are present. If abnormal cells are present, a sample, literally a piece of the abnormal region of the cervix, is pinched off and put in a container to send to the pathologist. Multiple biopsies may be necessary, depending on the appearance of the cervix. After all diseased areas are evaluated and biopsied, the bleeding is stopped. After a brief observation period, the patient is usually free to leave the clinic.

The pathologist reviews the samples and determines if they represent diseased portions of the cervix. If diseased portions are present, the pathologist will grade severity of the sample. Finally, the patient and doctor make a decision regarding how to proceed. Possible treatments include burning or freezing abnormal surface cells off the cervix (loop electrical excision procedure or cryosurgery), removal of a larger portion of the cervix (a cone biopsy), or, in severe cases, a full hysterectomy (removal of the cervix and uterus).

Usually, the cervix goes through multiple precancerous stages before actually becoming cervical cancer. These stages, in order of progression of disease, include:

- CIN[2] I
- CIN II
- CIN III

---

[2] *Cervical intraepithelial neoplasia.*

- CIS[3]
- Cervical cancer

A mild change in the cervix can be detected early because it usually takes years for the cervix to progress through the stages. Sometimes, however, in as little as one year, a woman can progress from normal to CIN III, CIS, or cervical cancer. This rapid progression is rare. The stages of the disease can generally be tracked before entering advance stages.

Some very dangerous HPV do cause a rapid transformation of the cervix from normal to cancerous. Currently, there is no way to know or tell in advance what kind of HPV your sexual partner may have, so it is important to limit sexual contact and always use latex or polyurethane condoms to reduce the risk as much as possible.

HPV comes in many strains or types, but two strains cause more than 70 percent of cervical cancers. A vaccine offers protection against these two types of HPV and provides some protection against genital warts strains of HPV. Discuss with your doctor if you are candidate for the HPV vaccine. Any female who is considering sexual activity or is already sexually active—a teenage sister, a daughter, a cousin, or even your recently divorced mom who is entering the dating scene after twenty-five years—should consider the HPV vaccination. Talk about it with the women and men in your life. You don't want your son, brother, father, cousin, best friend, or co-worker to either contract or pass on a virus that can potentially cause cancer. The danger is real, and we can't afford to keep quiet any longer. Attention, casual sex can now kill you in more ways than one!

---

[3] *Carcinoma in situ.*

The number of sexual partners matters. Yes, it only takes one encounter with an infected partner (usually male) to get a damaging HPV that causes cervical cancer, but you increase your risk of coming in contact with that virus with every new sexual partner. Each new partner brings you closer to coming in contact with the virus that causes cervical cancer. Gosh, that is scary! All persons must be very careful with chosen sexual partners.

Summary

It is crucial that we all take responsibility for our health. Get Pap smears, and get them on time, whatever that is for you. Talk to your doctor, and ask the questions. And if you do feel scared, let it be known. Make an appointment today.

Chapter Five

Mrs. Charleston

"What brings you in today, Mrs. Charleston?"

"I've been very tired. I've been dragging around the house and barely able to get out of bed for work for the past three or four days. I've been getting some neck and shoulder pain as well. Maybe I'm just stressed. Working too many hours at the firm."

Mrs. Charleston, a forty-four-year-old female, is a prominent attorney in the city. Truly a driven woman, she sometimes works fifteen hours a day when an important case is approaching. It is not uncommon for her to complain of fatigue while preparing for these cases. Mrs. Charleston's current case is a very high-profile one that has attracted more than its share of media attention. She received threats at home and at work and experienced difficulty sleeping during the two months she prepared for this case.

"Maybe you're just stressed, but let me decide if this is simply a case of an overworked executive," I tell her.

Mrs. Charleston and I have a mutually trusting relationship. She trusts my judgment and allows me to make decisions based upon medical knowledge. She doesn't try to bully me with her influence to prescribe certain medications or order tests. Our doctor-patient relationship is a professional one that, I believe, we both have come to appreciate over the past five years.

I ask further questions:

- Any changes in urination frequency and/or urgency?
- Any chest pain, chest pressure, or palpitations (rapid heartbeat)?
- Any numbness or tingling?
- Any bleeding like vaginal or rectal? Any dark stools?
- Any headaches, vision changes, or weakness?
- Sleep in an awkward position?
- Mood changes like depression or anxiety?

The answer to all these questions and more are either no or "I don't think so." After our discussion and interview, I examine her from head to toe. Her exam is completely normal. I have known her for five years, so I know her family history. Her mother was killed in a car accident at age thirty-eight, and her father died from lung cancer, probably related to a long smoking history. Mrs. Charleston is an only child with no children. She doesn't smoke, use drugs, or engage in excessive alcohol use.

Annually, she undergoes fasting blood work, which has been normal except slightly elevated cholesterol that she agreed to manage with a low cholesterol diet and exercise. Overall, she is a pretty healthy woman. So what about this fatigue and neck pain?

She's due for her annual physical, so I decide to get all fasting blood work. I also decide to get an electrocardiogram (EKG). All blood work and EKG return as normal. I allow Mrs. Charleston to go home with instructions to call if symptoms worsen or any new symptoms emerge.

The next day, Mrs. Charleston calls to get the results of her lab work. I relay the message that all is negative. She thanks me for being so complete in my evaluation. At the end of our conversation, Mrs. Charleston sighs deeply and states that she neglected to tell me that she had also been having some shortness of breath when climbing stairs. She figured she was just out of shape. She has been unable to work out lately—actually for the past couple years—because of her demanding job.

With this information, I make an appointment for her to have a stress test. We discover that Mrs. Charleston has heart disease that requires urgent treatment. She is doing well today because of the information she shared with me. Something she thought was unimportant became a pivotal point in the evaluation of her symptoms.

Heart Disease

Heart disease is a dangerous, sometimes silent, and insidious foe. It can be ridiculously simple to diagnose by its classic symptoms of chest pain with or without physical exertion, chest pressure, left arm pain, and shortness of breath. Or it can be amazingly sly and present with very vague symptoms of fatigue, neck pain, back pain, nausea and vomiting, or sweating.

Heart disease, the leading cause of death in the United States, is a blanket term to describe any of many diseases affecting the heart, including congestive heart failure, heart attack, cardiomyopathy, and so forth. Here, we will primarily focus on heart attack (myocardial infarction) resulting from coronary artery disease.

A blockage in the arteries (blood vessels) that serve the heart cause coronary artery disease. Thrombus (clotted platelets) and cholesterol deposits compose this blockage. Over time, the blockage gets wider and wider, then slows down, and eventually completely stops the blood flow through the arteries. When blood does not flow through the coronary (heart) arteries, the heart muscle cannot get the oxygen and nutrients necessary for life, and the cells served by this now-blocked artery begin to die. The death of the heart cells from lack of blood flow is called a heart attack. If detected early, stenting or a bypass can restart blood flow. Some heart muscle may be saved from complete death and resultant scar formation. Once scar forms, the heart muscle is permanently damaged and cannot be revived. If scar formation can be prevented and the disease is caught early, valuable heart function can be saved.

It is important to have an evaluation if you experience any symptoms, especially if you are a high-risk patient:

- At least forty-five years of age
- Male especially, but even female
- A smoker
- A diabetic
- Previous history of heart disease
- Hypertensive
- Hypercholesterolemia

Today, there are many ways to treat heart disease:

- One treatment includes using medication that will strengthen your heart along with cholesterol-lowering medications.
- Another option is stenting, the procedure in which the arteries of your heart are opened with a small tube to allow blood to flow freely through a blocked artery.

If heart disease cannot be treated with stenting, open heart surgery could be necessary. A vein from another part of the body, usually the leg or arm, serves as a bypass around the clotted, closed region of the artery.

Summary

In order to avoid heart disease, it is important to adopt a lifestyle which supports a healthy heart. Heart disease is caused by many of the same chronic conditions and unhealthy habits discussed in earlier chapters. Prevention of heart disease involves eating right, exercising, managing diabetes, blood pressure control, smoking cessation and taking aspirin if recommended. Speak with your doctor about your own personal risk factors for heart disease. It is important to review your family history, personal history, diabetes risk, blood pressure and cholesterol levels. Review your risk factors BEFORE you have symptoms. Even if you are not a high-risk patient, experiencing any warning symptoms warrants immediate medical attention. Speak with your doctor, ask the questions, and get the answers.

Chapter Six

Mr. Johansen

My next patient of the day, Mr. Johansen (Mr. Jo), who the staff knows well and likes, has not always been compliant or willing to take his medications. His blood pressure has never really been under control, only varying degrees of uncontrolled. He's here today because I asked him to follow up for discussion of his recent blood work.

Mr. Jo asks how my family and I are doing. He discusses his grandchildren and even takes out pictures for me. After our pleasantries, I tell Mr. Jo that I have a very important matter to discuss with him. This makes him uncomfortable. He slightly shifts in his chair and nervously glances around the room. I pull my chair in close to Mr. Jo and look him in the eye.

I state the obvious. "Mr. Jo, your blood pressure has not been controlled. We have been working on managing your blood pressure for years. You have seen the kidney doctors to help with your blood pressure. You have not always followed our advice with taking your medications."

Mr. Jo shifts nervously. "I know."

"Because of the years of uncontrolled blood pressure, your kidneys are now very damaged, and I suspect you will need dialysis very soon. We need to discuss how you feel about this and how this will be life changing for you. It is necessary that we have this conversation now because I'm unsure how long we have until your kidneys completely fail."

Mr. Jo drops his head and says nothing for many seconds. I allow the information to sink in and wait for him to speak.

"I don't want to be on dialysis. Isn't there some kind of medication I can take to make my kidneys well? You docs have a pill for everything. You must have something for this, don't you?"

"No, I don't. There is permanent damage to your kidneys from the high blood pressure. There is no medication to make them well again. The kidneys normally work to clean your blood, but, without functioning kidneys, a dialysis machine must filter and clean your blood. I'm sorry this is happening to you, and I wish there were something more to be done to save your kidneys. I wish we did not have to discuss this, but we need to prepare for beginning the process of dialysis. This will be life changing for you and your family. I'm here to help. Is there anyone—a family member, friend, or other person—with whom you'd wish for me to speak?"

Chronic Kidney Disease

Chronic kidney disease, generally a slowly progressive disease, may end in complete renal failure if circumstances surrounding the origination and advancement of disease are not addressed. Sometimes, even if the causes are effectively addressed, the disease can continue to progress.

The most common causes of chronic kidney disease are hypertension and diabetes, which cause progressive damage that can ultimately result in kidney (renal) failure. When kidneys fail, they become permanently damaged and are unable to sustain life.

Not only do the kidneys act as a filter for the blood, they also are important in hormone regulation. Once these organs are damaged, this can result in damage to the bones, electrolyte abnormalities (elevated potassium, phosphate, magnesium, and other abnormalities), protein loss, increased blood clotting ability, hypertension, and hypotension. Kidney failure patients also have an elevated risk of heart disease.

You must find out if you have kidney disease. Most people do not know they have it. Your doctor can easily test you for this disorder.

The best way to treat chronic kidney disease is to prevent it. Controlling hypertension and diabetes are essential to preventing kidney damage. It is especially important to avoid overexposure to certain medicines, especially anti-inflammatory medications (ibuprofen, naproxen, and naprosyn among others). You must minimize exposure to contrast-requiring tests like barium enema, upper gastrointestinal (GI) tests, computer tomography (CT) scans, and so forth. Consult with your physician to discuss other ways to maintain kidney function.

If you already suffer from chronic kidney disease, you must follow the above measures to possibly avoid end-stage renal disease and dialysis. If you have been lax about your health in the past, you must now get serious. You are to see your primary care physician and nephrologist as recommended. You are to follow all directions. It is the doctors' responsibility to communicate with each other; it is your responsibility to make sure you relay all information necessary between your doctors. For example, if your nephrologist orders lab work or other tests (ultrasounds, MRIs, and so forth), ask for a copy, and take it to your next visit with your primary care doctor. If your doctor has already seen and addressed these labs and tests, that's great, but, if not, you may be providing the valuable and necessary information that will assist with your own care.

If you already have end-stage renal disease and are on dialysis, you must care for your overall health, which includes getting enough rest, eating balanced meals and following the diet appropriate for you, keeping up with appointments, and, most importantly, taking all prescribed medications. Your medical doctor and nephrologist (kidney doctor) must first clear any vitamin supplements or herbs you are considering. Some supplements and over-the-counter medications can be harmful.

Summary

Once again, you are the important factor here in prevention and treatment. Speak with your doctor, ask the questions, and get the answers.

Chapter Seven

Mrs. Cooper

"I'm tired," says Mrs. Cooper, a stay-at-home mom for the past fifteen years, after I ask how she's feeling.

A former administrative assistant to the CEO of a Fortune 500 company, she left that position when she decided to get married and have children. She has a loving family, including her supportive husband, also a patient of mine. Mrs. Cooper usually comes in with her husband, but today is present alone. This is the sixth time I have seen Mrs. Cooper in as many months. Each time, she has a different complaint: headache, stomach pain, chest pain, and/or limb pain. She had an extensive workup of symptoms, sometimes with urgent specialist appointments. Each test and evaluation yielded only a clean bill of health for my fifty-three-year-old patient.

Today, she has the nonspecific complaint of fatigue. Mrs. Cooper states she is tired on most days and has difficulty getting out of bed, even after sleeping for ten hours during the night. She has been so fatigued that she now takes naps during the day as well. She has some difficulty concentrating and performing her usual responsibilities. She also mentions that she has not been to the stay-at-home mom group that she founded, for the past three months. Her fourteen-year-old son was recently diagnosed with diabetes, and she feels responsible for his poor diet and exercise habits.

She knows that some changes come with being fifty-three years old, but she has noticed lately that she isn't as quick thinking as she used to be. It takes longer for her to approach many problems subsequently longer to come to a solution. I notice she has gained about twenty-five pounds in the last six months. She says she is too tired to exercise and really doesn't like to do it anymore. This woman was an avid tennis player who competed in many local and statewide competitions. She had ruptured her Achilles tendon five years ago while playing tennis, recovered in record time (three weeks), and pressured the orthopedic surgeon to lift her restrictions early to play in upcoming tennis finals.

Mrs. Cooper continues to complain that the fatigue overwhelms her. I carefully review her medical record again for any clues. Mrs. Cooper doesn't complain unless she is really suffering, so even mild complaints from her are usually understated, which worsen my concern. I have known her for seven years, and this is the first time she has come to me with fatigue. I review, once again, all the testing I ordered over the past six months. All of Mrs. Cooper's random and fasting blood work is normal. She had a stress test and workup of her heart and heart function (in addition to cardiology consultation) secondary to recent complaints of chest pain, and they were normal. She had a CT scan of her brain and abdomen secondary to recent complaints of severe headache and abdominal pain. She had multiple x-rays secondary to shoulder, hip, knee, and neck pain. I could not find any cause for her fatigue.

What could be the source? I review multiple medical sources, books, and Web sites and ask colleagues, but I don't receive a solid answer. I become convinced of some type of rare disorder and want to begin ordering other tests that must be sent out of state to be processed. But before I embark on that path, I need more discussion with Mrs. Cooper. I want to be sure that I'm not missing any social concerns, namely, domestic abuse. I have not seen any bruising on Mrs. Cooper, but that doesn't mean abuse was not occurring. Because I also see her husband, I wonder if she felt uncomfortable initiating a discussion.

I ask about the possibility of being in a hurtful situation. I ask if she ever felt in danger or threatened by anyone, including her husband and children.

She looks at me with a puzzled glance. "No, I don't fear anyone. My husband has always put money into my account each month, outside of household bills, for me to use however I pleased. My God, if I felt I were being threatened or mistreated, I could have saved my money and left him years ago. I appreciate your concern, but, no, I'm not being abused verbally, physically, or otherwise."

Her response sufficiently satisfies me, but I'm still uncertain about the overwhelming fatigue. Then I finally get it. Why haven't I caught this before now? "Have you been feeling depressed or down lately?"

Major Depression

We all go through times of feeling down and having difficulty with traumatic events. That is not major depression. Major depression is an overwhelming feeling of depressed mood with the symptoms listed below for two or more weeks that impact your day-to-day functioning. If you have any questions about the possibility of depression and would like clarification or further discussion, make an appointment and talk to your doctor. Maybe you are right and that fatigue really is from another medical condition, possibly something as simple as anemia (low blood count), but what if you are wrong? You could spend years suffering without treatment.

Major depression is generally insidious in its onset. It can begin suddenly because of traumatic events or slowly as events and things that used to be important slowly slip into the background of life. Both men and women may:

- Begin sleeping more or less
- Lose or gain weight
- Have symptoms of decreased concentration, guilty feelings, and anhedonia (the sense no longer enjoying activities that you used to enjoy)
- Become slow-moving or slow-thinking
- Have suicidal thoughts with or without a depressed mood and/or crying spells

You can't just "shake off" major depression. It is an imbalance of brain chemicals called neurotransmitters caused by multiple factors working together. You may have negative or depressing situations with or without genetic inheritance factors. These things work together to cause a deficiency in neurotransmitters, usually serotonin and norepinephrine. The end result—regardless of the route to get there—is still the same, and that is major depression.

It is not my intention to suggest that every person with more than one symptom or medical complaint has major depression. In fact, many chronic diseases mimic depression or can make depression worse or even precipitate depression in a person already prone to it. All complaints must be addressed appropriately.

Many people may not realize that major depression is the cause of their complaints and may resist the diagnosis, even in light of overwhelming evidence. If you are concerned and do not feel your symptoms are from depression but your doctor feels otherwise, it is okay to have an evaluation from a licensed psychiatrist or another physician for a second opinion. If the diagnosis is confirmed, it is important you receive medical attention.

Medication may be required to restore chemical balance in the brain. Some patients may require counseling and psychotherapy with or without medication. Each patient is unique, but a treatment plan is necessary.

If you require medication, you may have to take the prescription for at least six weeks before you see an effect. You may have side effects from one, two, or more antidepressant medications and may have to have them switched until the right one(s) is found for you. It is important to have patience. This disease takes time to effectively treat, but the treatments are helpful and successful.

If you and your doctor decide you need medication, take it. Take it as prescribed to receive the maximum benefit. Report all side effects you experience, especially those that make you feel like stopping the medication (such as sexual dysfunction, weight gain, headache, abdominal pain, or nausea). These side effects may be completely resolved or at least greatly minimized with just a few small modifications, such as adjusting or changing your medications completely. You may need to be on your medication a couple months, a year, two years, or a lifetime.

Patients with depression must keep themselves generally healthy with a balanced diet and exercise. Extra weight and inactivity can worsen symptoms, and, if there are any chronic diseases, it is important they are optimally controlled. For example, uncontrolled diabetes can worsen depression symptoms. If you are depressed and forgetful, forgetting to take your medications or taking them incorrectly because of difficulty concentrating or following the treatment regimen, you will not control the chronic disease, and your overall health will worsen along with the depression.

The most dangerous consequences of depression, of course, are suicide and homicide. If you feel you may hurt yourself or someone else, reach out to your primary doctor, an emergency room, or urgent-care facility for help. You need treatment now. Religious and spiritual practices are important, but may not be enough alone. Many people in religious groups do not receive the help they need and deserve because of depression being viewed as a kind of spiritual attack. Medical science notes the existence of chemical imbalance in the brain. Major depression is real, just as diabetes and high blood pressure are real. It is unlikely you will be rid of it without some form of treatment.

Summary

Whatever your treatment regimen, follow it. You deserve to be an active participant in your life. Once again, you are the important factor here in prevention and treatment. Speak with your doctor, ask the questions, and get the answers. You are starting to get the message!

Chapter Eight

Mr. Tyson

"Doc, I want a physical. I came in today because I need a physical."

"Are you having any concerns, Mr. Tyson?"

"Nope, just haven't had a physical in a while, and I need one."

"Okay, well, let's get started."

I ask about everything: new medications, other past medical history, surgeries, and social history including alcohol and tobacco use. The next part in the medical history reviews systems, during which I ask patients about everything head to toe, including weight loss or gain, fever or chills, headaches, and abdominal pain. At that point, Mr. Tyson becomes fidgety. He avoids eye contact and glances at the ground. Mr. Tyson has been a patient of mine for five years, and I have never seen him behave in this manner.

I acknowledge his obvious apprehension. "You seem to be uncomfortable. Is there anything I'm asking or doing that is making you feel uncomfortable?"

He quickly answers, "No."

I proceed with my interview. Finally, I progress to the questioning regarding sexual function. I have asked the same questions at every physical before this and assume his answers would be the same.

"Are you having any difficulties with sexual functioning? Are you having trouble getting or keeping an erection?"

"Huh?"

Knowing full well this mechanical engineer understood my question, I question him in a different manner. "Are you having any difficulty with performing sexually?"

"Uh … yeah … Uh … I, uh, haven't been, you know, as strong or able to perform as long lately. I wanted to talk to you about this before, but I couldn't. It has been going on for about six months. I've been avoiding my wife, and I know she thinks I'm cheating. Don't take it personally, but, you being a woman and all, I didn't feel comfortable talking to you."

"Don't think of me as a woman. I'm a doctor. If you want to talk to one of my male colleagues about this, I'm perfectly happy to refer you to another physician."

"God, no! I could barely get it out to you; I don't want to talk to anyone else. What can you do to help me?"

Erectile Dysfunction

Erectile dysfunction is difficulty with getting or keeping an erection. This can occur for many different reasons, including:

- Prescription medications
- Chronic diseases such as hypertension or diabetes
- Prostate abnormalities
- Elevated cholesterol
- Emotional issues and concerns
- Performance anxiety
- Depression
- Fatigue
- Over-the-counter medications
- Illegal drugs (including marijuana) and/or legal drugs (alcohol and cigarettes)

Your physician must investigate erectile dysfunction thoroughly, which is why it is essential that you find someone with whom you feel comfortable. Occasionally, erectile dysfunction is the initial symptom diagnosing multi-organ arterial disease (hypertension, diabetes, and so forth). Early evaluation and management may prevent other consequences like stroke, limb amputation, kidney failure (dialysis), and heart attack.

Be forewarned that the office visit probably will require a prostate exam. This is not something to get anxious about. During the prostate exam, the doctor will insert a gloved finger into the anus to feel the small gland just inside the rectum. The physician will be feeling for an enlarged prostate, masses, hardness, or tenderness. The patient may also need a blood test called a prostate specific antigen or PSA. A discussion must occur with the doctor to determine if a PSA is necessary.

If you are a woman, it is important for you to take control of this for the man (husband, boyfriend, or close friend) in your life. If he doesn't feel comfortable addressing this topic, ask to accompany him to his next doctor visit, raise the topic to the physician, and then excuse yourself from the exam room so he may have a confidential conversation regarding sexual health. Another option is to send a brief note with the patient outlining the major concerns. The physician cannot give you any information without the written consent of the patient, but you can relay all the information that you feel is necessary to the patient's doctor. You have a responsibility to be a good wife or friend to the males in your life and to assist if you know they won't get help on their own.

The woman of the house is generally the primary health-care decision-maker. Ladies, it is important that you find a doctor with whom the man in your life feels comfortable. This may mean you and your significant other have two different physicians. That is okay. Some women are only comfortable with female physicians; some men may only feel comfortable with male physicians. I don't care if you feel comfortable only with a purple people eater for a physician. You need to find Dr. People Eater to be your primary care doctor. This is a sensitive issue, and many men will get the treatment necessary only if they can be completely honest and open with all concerns and symptoms.

Summary

It may take a couple visits before a patient feels comfortable discussing such private matters. It sometimes takes years to form close friendships, and many people still do not have anyone they can tell intimate secrets. Erectile dysfunction is not to be taken lightly because of the medical and emotional implications involved. This cannot wait more than three or four visits (less than three to six months). Get as comfortable as you can, give yourself time to find a physician you trust, and figure out a way to address this major concern. It may be more than just a performance issue. Once again, you are the important factor here in prevention and treatment. Speak with your doctor, ask the questions, and get the answers. I think you get it now!

Chapter Nine

The Physical Examination

This chapter is a little bit different. In it, I discuss what is necessary at every age during a routine physical examination. What should you be looking for? What kind of general testing should be ordered and so forth? The following will be an age-by-age breakdown of what is to be expected:

Newborn	Birth to one month	<ul><li>Length</li><li>Weight</li><li>Heart and breathing rate</li><li>Head circumference (distance around the head)</li><li>Head-to-toe examination</li><li>Discussion regarding safety</li><li>Social concerns like siblings, daytime care, and other family members in the home</li><li>Feeding and nutrition</li><li>Bowel movements</li><li>Sleep patterns</li><li>Anticipatory guidance</li><li>Umbilicus</li><li>Circumcision (males)</li><li>Immunizations</li></ul>

Infants	One month to one year	<ul><li>Height</li><li>Weight</li><li>Heart and breathing rates</li><li>Head circumference</li><li>Head-to-toe examination</li><li>Discussion of immunizations</li><li>Discussion regarding safety and anticipatory guidance (what is expected in your baby's development)</li><li>Sleep patterns</li><li>Appropriate toys</li><li>Social concerns like siblings, daytime care, and other family members in the home</li><li>Feeding and nutrition</li><li>Bowel movements</li></ul>
Toddler	One to four years	<ul><li>Height</li><li>Weight</li><li>Vision and hearing screening</li><li>Blood pressure</li><li>Heart and breathing rate</li><li>Body mass index (BMI)</li><li>Head-to-toe examination</li><li>Social concerns like siblings, daytime care, and other family members in the home</li><li>Discussions about what to expect and how to handle tantrums and picky eaters</li></ul>

		• Safety • Discussion of language development and play • Sleep patterns • Appropriate toys • TV/media time • Anticipatory guidance • Discussion of feeding and nutrition • Bowel movements • Immunizations
Young child	Four to seven years	• Height • Weight • Vision and hearing screening • Blood pressure • Heart and breathing rate • BMI • Head-to-toe examination • Discussion of immunizations • Safety • Language development and play • Sleep patterns • Appropriate toys • TV/media time • School performance • Anticipatory guidance • Feeding and nutrition
Child and	Eight to twelve	**Male**

preteen	years	<ul><li>Height</li><li>Weight</li><li>Vision and hearing screening</li><li>Blood pressure</li><li>Heart and breathing rate</li><li>BMI</li><li>Head-to-toe examination</li><li>Nutrition</li><li>Accident prevention</li><li>Peer pressure</li><li>Expected body changes during the coming years</li><li>Discussion of immunizations</li><li>Anticipatory guidance and school performance</li><li>Sleep patterns</li><li>TV/media time</li><li>Internet safety</li><li>Depression screening</li><li>Discussion about sex—what it is and is not—before the hormones begin to rage</li></ul>
		Female <ul><li>Height</li><li>Weight</li><li>Vision and hearing screening</li><li>Blood pressure</li><li>Heart and breathing rate</li><li>BMI</li></ul>

		• Head-to-toe examination • Nutrition • Accident prevention • Peer pressure • Discussion of expected body changes during the coming years • Immunizations • Anticipatory guidance and school performance • Sleep patterns • TV/media time • Internet safety • Depression screening • Discussion of menstruation and sex—what it is and is not—before the hormones begin to rage
Teen	Thirteen to eighteen years	**Male** • Height • Weight • Vision and hearing screening • Blood pressure • Heart and breathing rate • BMI • Head-to-toe examination • Nutrition including, but not limited to, obesity, eating disorders, proper diet, and use of supplements

		• Driving and accidents
		• Peer pressure and responsible choices
		• Safest sexual behaviors
		• Discussion of organized sports and possibilities of joint and limb accidents
		• Wearing appropriate protection and gear during sports
		• Avoidance of legal (tobacco and alcohol) and illegal drugs and weapons
		• Discussion of immunizations
		• Sleep patterns
		• TV/media time
		• Internet safety
		• Depression screening
		• School performance and future planning
		Female • Height • Weight • Vision and hearing screening • Blood pressure • Heart and breathing rate • BMI • Head-to-toe examination • Nutrition including, but not limited to, obesity, eating disorders, proper diet, and use

		of supplements • Peer pressure and responsible choices • Driving and accidents • Safest sexual behaviors • Discussion of family planning, menses, and loss of menses with intense physical activity • Avoidance of legal (tobacco and alcohol) and illegal drugs and weapons • Discussion of immunizations • Depression screening • School performance • Sleep patterns • TV/media time • Internet safety • Future planning
Adult	Childbearing years, nineteen to thirty-nine years	**Male** • Routine physicals every one to two years • Vital signs • Height • Weight • Blood pressure • Heart and breathing rate • BMI • Head-to-toe examination • Fasting blood work (possibly

		urine) • Nutrition • Exercise and stress reduction • Depression screening • Substance abuse screening • Discussion of immunizations • Occupational counseling
		Female • Routine physicals every one to two years • Discussion and family planning information • Height • Weight • Blood pressure • Heart and breathing rate • BMI • Head-to-toe examination • Fasting blood work (possibly urine) • Pap smears every one to two years • Nutrition • Exercise and stress reduction • Depression screening • Substance abuse screening • Discussion of immunizations • Occupational hazards counseling

Adult	Forty to sixty-four years	**Male** • Routine physical yearly • Height • Weight • Blood pressure • Heart and breathing rate • BMI • Head-to-toe examination • Prostate screening for high-risk groups at forty to fifty years • Colon cancer screening at fifty years old • Lab work including fasting blood work and urine testing • Discussion regarding exercise, nutrition, and maintaining current activity level or better • Stress reduction • Depression screening • Substance abuse screening • Discussion of immunizations • Discussion of any changes in sexual function or activity
		Female • Routine physical yearly • Height • Weight • Blood pressure • Heart and breathing rate • BMI

		• Head-to-toe examination • Pap smear • Annual mammogram • Colon cancer screening at fifty years old • Lab work including fasting blood work and urine testing • Discussion regarding exercise, nutrition, and maintaining current activity level or better • Stress reduction • Depression screening • Substance abuse screening • Discussion of immunizations • Discussion of any changes in sexual function or activity
Seniors	Sixty-five years and up	**Male** • Yearly physical with check-ups in between as needed for chronic disorders • Height • Weight • Blood pressure • Heart and breathing rate • BMI • Head-to-toe examination • Continue with colon and prostate cancer screening • Discussion regarding maintaining activity, nutrition, and proper exercise

		<ul><li>Lab work including fasting blood work and urine testing</li><li>Discussion of immunizations</li><li>Discussion of any changes in sexual function or activity</li><li>Depression screening</li><li>Substance abuse screening</li><li>Memory discussion and possible testing</li><li>Hearing screen</li><li>Home risk assessment for safety</li></ul>
		Female <ul><li>Yearly physical with check-ups in between as needed for chronic disorders</li><li>Height</li><li>Weight</li><li>Blood pressure</li><li>Heart and breathing rate</li><li>BMI</li><li>Head-to-toe examination</li><li>Continue with breast cancer screening</li><li>Cervical cancer screening</li><li>Colon cancer screening</li><li>Bone density testing</li><li>Nutrition and proper exercise</li><li>Lab work including fasting blood work and urine testing</li><li>Discussion of immunizations</li><li>Discussion of any changes in</li></ul>

		sexual function or activity
		• Depression screening
		• Substance abuse screening
		• Memory discussion and possible testing
		• Hearing screen
		• Home risk assessment for safety

Epilogue

I have written this book for patients everywhere, and I hope this information will help all it touches. It really is possible for you to have the best care available. You must start with a good foundation, a good primary care physician. The definition of a good doctor is one with whom you relate well. It is possible and necessary to have your individual needs met. This book provides the beginning and the framework as just a start. The real work still lies ahead. If you begin building your relationship with your doctor on a solid foundation, it is more likely to be stable and strong when or if you should ever need serious medical counsel, a health-care advocate, or simply a shoulder to cry on.

Your doctor can be an indispensable asset. He or she can be amazingly helpful. It is in your best interest to choose a doctor with whom you can develop a relationship with over time. You can do it. I know you can.

Throughout this book (in different forms), you have seen, "You are the important factor here in prevention and treatment. Speak with your doctor, ask the questions, and get the answers." This is truly the key. Open the lock to good health. Take care of yourself and those you love.

Wishing You Great Health,

Dr. Frances Anderson-Hewitt

Resources

Your trusted primary care physician, friends and family members in the medical field, your local pharmacist and pharmacy staff, your local county health department, your local hospital Web sites, your local medical school Web sites, and general Web sites

Licensing Web sites	*your state*.gov/licensing for healthcare professionals
Board certification documentation Web sites	www.abms.org
WebMD	www.webmd.com
National Institutes of Health	www.nih.gov
Centers for Disease Control and Prevention (CDC)	www.cdc.gov
Women's and Men's Health	www.womenshealth.gov
Diabetes	
American Diabetes Association	www.diabetes.org
American Academy of Family Physicians	www.familydoctor.org
Hypertension	
National Heart Lung and Blood	www.nhlbi.nih.gov

Institute	
American Academy of Family Physicians	www.familydoctor.org

Sexually Transmitted Disease/Infection

Centers for Disease Control and Prevention (CDC)	www.cdc.gov
American Academy of Family Physicians	www.familydoctor.org

Cervical Cancer

National Cancer Institute	www.cancer.gov
American Cancer Society	www.cancer.org

Heart Disease

American Heart Association	www.americanheart.org
American Academy of Family Physicians	www.familydoctor.org

Chronic Kidney Disease

American Kidney Fund	www.kidneyfund.org
National Kidney Foundation	www.kidney.org

Depression

National Institute of Mental Health	www.nimh.nih.gov
American Academy of Family Physicians	www.familydoctor.org
American Psychiatric	www.psych.org

Association	
American Psychological Association	www.apa.org
Center for Mental Health Services	www.mentalhealth.org

Erectile Dysfunction

American Academy of Urology	www.auanet.org
American Academy of Family Physicians	www.familydoctor.org

Preventive Screening

American Academy of Family Physicians	www.familydoctor.org
US Preventive Services Task Force	www.uspreventiveservicestaskforce.org
US Department of Health and Human Services	www.hhs.gov

About the Author

Dr. Frances Anderson-Hewitt is a board-certified family physician practicing in Detroit, Michigan. She was born and raised in Chicago, Illinois, where she completed elementary and high school. She began her higher education at the University of Illinois at Urbana-Champaign and graduated with a bachelor's degree in biology. She then completed medical school training at The Ohio State University College of Medicine. She then secured a residency position at Henry Ford Hospital in Detroit and currently serves as faculty and senior staff physician for the Department of Family Medicine at Henry Ford Hospital, where she is actively involved in her own clinical practice and teaches medical students and resident physicians.

Dr. Anderson-Hewitt now resides in the Detroit metropolitan area with her family. She is involved in community activities through Hope United Methodist Church in Southfield, Michigan, and other local churches and community centers in the area. Dr. Anderson-Hewitt is a Diamond Life member of Delta Sigma Theta Sorority, Inc.

www.ingramcontent.com/pod-product-compliance
Lightning Source LLC
Chambersburg PA
CBHW022105170526
45157CB00004B/1492